The Steel Band

John Bartholomew

Oxford University Press
Music Department, Ely House,
37 Dover Street, London W1X 4AH

Contents

Introduction

The Steel Band was invented around the time of World War Two. Its home is Trinidad: its roots are in Africa, where ancestors of the Trinidadians came from.

The steel banders call the music **Pan**. The oil drums on which it is played are 'pans'. Trinidadians have always believed that music could and should be made with whatever came to hand. Bamboo sticks, bottles and spoons, car parts such as hubcaps and brake drums, dustbins and biscuit tins; all these modest items have a place in the rich cross-rhythms of the country. Nothing could be more natural than that with the discovery of oil in the island, the oil drum should take its place as the basic material for an instrument.

The Trinidadians claim that the steel drum is the only musical instrument invented this century. (The synthesizer is not really a new instrument—it is an adaptation of a keyboard instrument using electronics, and is not therefore 'new'.) In the lifetime of the panmen, Pan has captured the imagination of musicians all over the world. It is an amazing sound, capable of doing justice to the typical calypso of Trinidad's Carnival, and equally to classical pieces such as the *William Tell* overture.

Steel bands are quite common now in Britain, especially in schools. Many British people are joining the fellowship of Pan, and long to know more about the story and science of steel drums. This book provides information about origins, manufacture, musical arrangements, playing techniques, and many other facets of the steel band. All this is set against the background of Trinidad. Pan may be worldwide now—but to understand it, you have to understand its home country first.

1 Trinidad and the Caribbean

On October 12th, 1492, Christopher Columbus landed at San Salvador in the Bahamas: thus began his discovery of the New World.

He believed that his voyage westward would bring him to India, and therefore named the Caribbean Islands 'The West Indies'. He called the people he met 'Indians'—the name that has stuck with America's original natives ever since. He found first the peaceful Arawak people; later on other islands, the very different and hostile Caribs. The latter, known in Spanish as 'Caribal', were eaters of human flesh and have given the word 'Cannibal' to our language. Both Carib and Arawak died out almost completely within a few years of the arrival of Europeans.

Columbus made four voyages in all, and came to realize that he had found not a continent, but a string of islands. South of the Bahamas, two other groups, the Greater and the Lesser Antilles, complete the chain. The Caribbean chain stretches in a huge bow 3,000 km from Cuba, off Florida, USA to Aruba off the Venezuelan coast. The islands surround the very deep Caribbean Sea which reaches down, in places, nearly 1,000 metres.

There is no typical Caribbean island. Different people have settled in each of them and made it entirely different from its neighbours. Wars have been fought over them. Five hundred years have seen many conquerors come and go. You can hear English, Spanish, Portuguese, Dutch and French spoken. Treasure hunters, pirates, planters and slaves have all been absorbed into the populations.

The southernmost of the islands belongs more to the South American mainland than to the Antilles. It is not surrounded by deep sea; it is separated from the coast of Venezuela by a few miles of fairly shallow water. This is **Trinidad**.

(title page) A market scene on the West Indian island of Grenada. The lady in the centre is selling cornmeal with a pint measure

◀ A sandy beach in Northern Trinidad

History of Trinidad

The Caribs lived in Trinidad undisturbed for perhaps five thousand years, until Columbus on his third voyage landed here and found their villages.

Although Columbus reported his discovery, it was thirty years before the Spanish started to come to Trinidad, and when they did so, it was not to form a colony. Their search was for the fabulous gold of **El Dorado**; Trinidad was a convenient base camp for their exploration of the Orinoco River. The first Spanish settlement was built in 1592 at San Jose (now St Joseph), and the colonists made their fortunes growing cocoa. Towards the end of the 18th century, however, an insect blight crippled the cocoa industry, and almost all the colonists were forced to leave. In 1773 the entire population of Trinidad had declined to about two thousand. Neighbouring Tobago, a fraction of Trinidad's size, had six or seven times as many people.

At about this time, at the other end of the Caribbean, the slaves in French-owned Haiti revolted, and drove out their French masters. Many of these found a home in Trinidad, and the population surged again. The French re-established the cocoa plantations, and also grew coffee and sugar. Sugar was to become the centre of Trinidad's economy.

In 1797 the short period of French domination came to an end. The British fleet captured Trinidad, which became a British possession until independence in 1962. Under the British the sugar plantations flourished, and the key to their success was the use of **slave labour**. The slave trade brought a new population to Trinidad. The descendants of the Africans who were dragged so unwillingly from their villages, and transported in the most inhuman conditions across the sea to a strange new world, form half of the island's present peoples. When the slaves

▲ African slaves waiting to be auctioned

were freed in 1834, they mostly chose to leave the plantations. The sugar-growers were faced with a labour shortage which threatened disaster. Workers from India were invited to fill this gap, and they came in thousands. In exchange for five years' work, they were each granted five acres of land. The Indians of Trinidad, who now make up a third of the population, have dominated the sugar industry from that time.

Carib, Spanish, French, British, African, Indian: each of these races has given something to the history of Trinidad. English is the official language, but you can also hear dialect Spanish, French, and Hindi.

In the present century, oil has been discovered. It has made Trinidad one of the most prosperous islands in the Caribbean. It is refined in Trinidad, and the refineries of course (because of the oil drums) are vital to the subject of this book. Oil is now the most important export, though the traditional crops are also still major exports. Industry, too has an increasing part to play. There is shipbuilding, a steelworks, and many factories manufacturing goods for export. From the Pitch Lake in the south of the island, asphalt is taken, and used on roads all over the world.

Many Trinidadians work in the tourist industry. It is the country's third biggest money earner. The beaches, especially on Tobago, are one of the attractions. The carnival music—calypso and steel band especially—is without doubt one of the others.

Car assembly plant on the outskirts of Port of Spain ▶

▼An offshore drilling rig

Regions of Trinidad

Three highland ranges dominate Trinidad. **The Southern Highland** includes three peaks, the Trinity Hills. These were Columbus' first sighting of the island and caused him to name it Isla de la Trinidad. **The Central Highlands** are a centre of cocoa growing, and also the site of the Naval Reservoir, the largest in the Caribbean. **The Northern Range**, three times as high as the others, almost cuts off the coast from the rest of the island. Few roads cross it. It is scored by deep valleys, and covered by jungle in many parts.

Along the foot of these mountains, to the south, is a line of towns including **Port of Spain**, the capital and main port. The remainder of the island is lowland. **The Caroni plain**, between the Northern and Central mountains, includes some swamp and grasslands, but is mainly fertile. The main sugar plantations are found at the western end of this plain.

The Southern (Nariva) Plain, and the **South West Peninsular** together with the Southern Highland include the inland oilfields, more cocoa plantations, and a lot of forest. The most productive oilfields are those offshore in the **Gulf of Paria**. The oil here is fairly easy to raise as the gulf has a maximum depth of only 27 metres. (The British oilfields in the North Sea are commonly under 150 metres of water.)

Tobago, an island 32 km north-east of Trinidad, is part of the same country. It is mainly mountainous and has through its history depended mainly on sugar plantations. It is thought by many to be the country's most beautiful region, and nowadays attracts many tourists.

(above) Citrus groves in the Santa Cruz valley, Northern Highlands. The trees in the bottom left-hand corner are banana trees

(below) Reaping cane on a sugar plantation

8

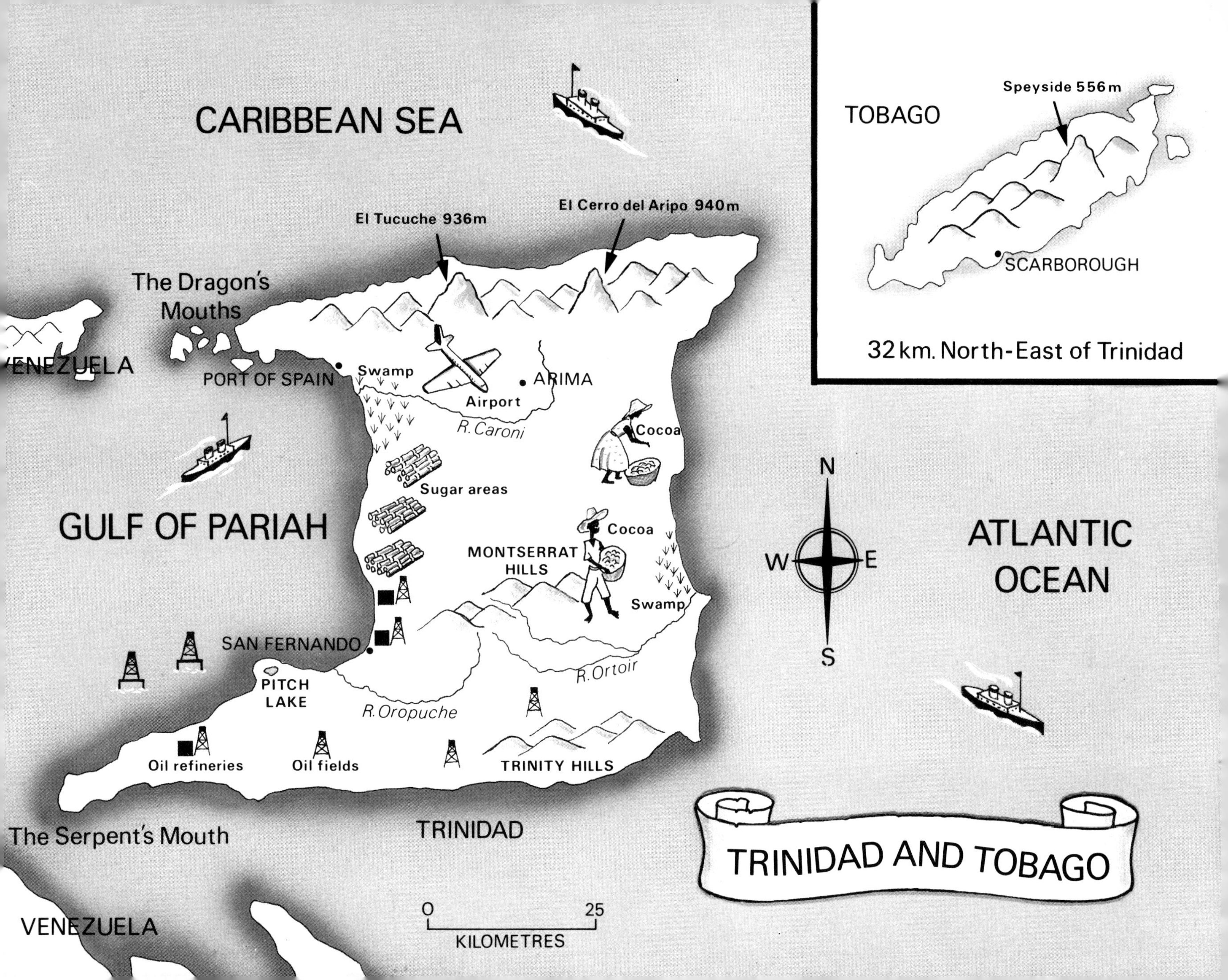

CARIBBEAN SEA

El Tucuche 936m

El Cerro del Aripo 940m

The Dragon's
Mouths

VENEZUELA

PORT OF SPAIN

Swamp

Airport

ARIMA

R. Caroni

Cocoa

GULF OF PARIAH

Sugar areas

Cocoa

MONTSERRAT
HILLS

Swamp

SAN FERNANDO

R. Ortoir

PITCH
LAKE

R. Oropuche

Oil refineries

Oil fields

TRINITY HILLS

The Serpent's Mouth

TRINIDAD

VENEZUELA

0   25
KILOMETRES

N
W   E
S

ATLANTIC
OCEAN

TOBAGO

Speyside 556m

SCARBOROUGH

32 km. North-East of Trinidad

TRINIDAD AND TOBAGO

Life in Trinidad

Port of Spain is a modern, planned city. It has grown up because this is the one place where there is a deep-water harbour that is also sheltered from the prevailing winds. The mountains to the north, and the sea and swamp to the south prevent it from spreading, except eastward. In this direction new developments sprawl along the foot of the mountains along the main road.

The oldest part of the city is also the centre. Narrow streets cluster around Woodford Square, with its cathedral and government buildings. Shoppers crowd along the famous Frederick Street.

Smart houses on the hillsides overlook the town, its harbour, and the sea. By contrast, the edge of the city near the swamp is a very poor district.

The Queen's Park Savannah is famous as a cricket ground and racecourse. It is also remarkable for its **food market**, a place that tells us a lot about the people of Trinidad. Nowhere more than in their eating habits are the varied origins of the people revealed. Of course, each community has 'borrowed' favourite dishes from the others. Spicy cooking, for example, has always been the favourite in Trinidad, and when the Indians arrived to work in the plantations, they brought with them a new range of spices and curries that have become some of the national dishes. They also brought rice, which is cultivated in the rainy season, and is now a basic food of the country. 'Souse', made from the meat of pigs' trotters in a cool dressing is made in

nearly every house. Seafood, from oysters (a small mangrove swamp variety) to shark steaks, are popular. Citrus fruits— lemons, limes, oranges and grapefruits—coconuts, mangoes, and guavas are all grown in Trinidad, and are part of the national diet. The main vegetables include yams, sweet potatoes and peppers. Rum is drunk everywhere, of course (sugar is its main ingredient), but some soft drinks are also a speciality— ginger beer, for example, which was first made for the plantation slaves who were not allowed rum. Angostura bitters is made only in Trinidad, to a closely-guarded secret recipe.

Trinidad is a tropical country, and like the rest of the Caribbean countries, has a constantly warm temperature. The average scarcely varies between 25° and 27°. The greatest contrast of climate comes in the rainfall. There is a rainy season from June to December, and a marked difference between the east of the island (which has much more rain, especially on the mountain section in the north), and the west. When it rains in Trinidad, it rains hard, and people are driven indoors while the shower lasts. Unlike the rest of the West Indies, Trinidad seldom sees a hurricane—it lies to the south of the main hurricane path.

Nearly half of the country lies under **forest** of various types. Pitch pine and teak are grown abundantly. Some of the forest on the northern range is rain forest, and makes for evergreen jungle scenery.

There is a varied **wild life**, too, with alligators, snakes, and monkeys among the species found. The striking image is colour. Marvellously brilliant colours are to be found on flowers, butterflies, and birds. The scarlet ibis is the national bird. And the original name of Trinidad—'Iere' as the Arawak people called it—means 'Land of the Humming Bird'.

Trinidadian market scene.
How many different fruits and vegetables can you spot? ▶

◀ View of the roof tops of Port of Spain.
On the left are the chimneys of one of the power stations

Questions

1 What are the three main groups of islands in the Caribbean called?
2 The original peoples of the American continents are sometimes referred to as 'Amerindians'. Can you name the two Amerindian races that were found in the Caribbean?
3 Which is the largest Caribbean island?
4 Place these settlers and inhabitants of Trinidad in the correct order of their arrival:
  British, Spanish, Carib, Indian, African, French.
5 What was mainly produced in Trinidad:
  a) By the Spanish?
  b) By the British after 1797?
  c) In the twentieth century?
6 What is the highest mountain in Trinidad?
7 Why did Columbus give Trinidad that name?
  What was the island called before that?

Projects

1 Use the following information to make graphs about the people of Trinidad.

Ethnic Origin		Religions in Trinidad	
African descent	42·8%	Christian:	
Indian descent	40·1%	Roman Catholic	36%
Mixed race	14·2%	Anglican	21%
White	1·2%	Other	13%
Chinese	0·9%	Hindu	23%
Other (inc. Amer-	0·8%	Muslim	6%
indian)		Others	1%

2 Make a relief map of Trinidad. You will need a firm base board, and papier-maché pulp made in a bucket from strips of newspaper mixed with glue. Using the map on page 9, or better still, a larger scale one from a good atlas, mould the shape of the island showing the high Northern range, the lower Central and Southern ranges, and the plains in between. Finish off by pasting over larger strips of paper, and leave it to dry. (IMPORTANT: wash out your bucket before the pulp dries.) When dry, paint to show sea, swamps, towns, etc. Miniature balsa wood oil rigs, etc. can be used to show other important features.

3 Most Trinidadian families make ginger beer. It takes some time, but is worth it. Some of the ingredients may be hard to find in this country, and I have given substitutes where necessary.
Screwtop bottles
50 grams of green ginger (or four teaspoons of ground ginger)
Juice and rind of two limes (lemons are nearly as good)
$\frac{1}{2}$ kg of caster sugar
4 litres of boiling water
25 grams yeast (use half this quantity if using dried yeast)
$\frac{1}{2}$ teaspoon of cream of tartar
Wash and pound the ginger to release the juices, and add the juice and rind of the limes or lemons. Pour on the boiling water. Stir in sugar and cream of tartar. Leave to cool, stirring now and then. When it is LUKEWARM add the yeast, dissolved in a little warm water. Cover and allow to stand overnight. Bottle and keep about a week before drinking. After about three days you would be advised to release some of the pressure by loosening and re-tightening the cap. This reduces the danger of the bottle exploding when you open it for drinking.

4 Find out more about the slave trade. There are many books about the subject.

The people of the West Indies have a reputation for loving rhythm. At **Carnival** especially, the streets pulse as the revellers make their way through the town. The beat is infectious, and those who come to watch are soon swept up in the dancing. Nowadays, it is the **steel band** that provides the characteristic carnival beat. Before steel bands were thought of, there had to be other music to stimulate the crowds.

What—or when—is Carnival? For the people of Trinidad it has always been a great celebration ending on the stroke of midnight on 'Mardi Gras'—Shrove Tuesday, the day before Ash Wednesday, the first day of Lent.

Carnival has changed its nature as the years have gone by. In the late 18th century, it was a high society affair, brought to Trinidad by the French settlers of that time. It was a time when the rich would visit each other, showing off in the streets with their fancy dress costumes. Masked balls (dances where the guests wore fancy dress and masks) were the highlight of the celebration. For the slaves, the great celebration was *Cannes Brulées*—the burning of the sugar cane stubble. This involved a torchlight procession. Then in 1833, the slaves became free, and thousands of Africans began to make Carnival their own. The Latin-American sounds of the dance bands gave way to a much more basic street rhythm, played mostly on drums, accompanied by bottles and spoons and whatever else came to hand.

The drumming, and dancing crowds of Africans, worried the authorities. Perhaps they thought it was too similar to the war dancing of their jungle ancestors to be safe. Whatever the reason, drums were banned from the carnival.

(title page) A pan player taking a rest from the carnival festivities

◄ A typical carnival costume: a gigantic headress in black, yellow, white, blue and red

Tamboo Bamboo

Drums or no drums, it was not long before the people took rhythm on to the streets once more. Lengths of cane and bamboo, miscellaneous pieces of wood and beaters, were the basis of the new bands. **Tamboo Bamboo** was the name given to them. It comes from the French word 'tambour' meaning drum.

The variety of sound from Tamboo Bamboo bands was amazing. The bamboo was prepared into all sorts of different forms. It was struck on the ground, banged together, notched and scraped. Different lengths of wood gave different pitches. Tamboo Bamboo bands took over the carnival, and the celebrations were as wild and noisy as ever.

In Trinidad there is a popular sport of 'stick fighting'. Like boxing, this sport is considered an art, controlled by strict rules. Also like boxing, it has its origins in crude, ferocious and often brutal fighting.

▲ A tamboo bamboo band; the players beat sticks on hollowed out bamboo canes

◀ Stick fighting is a popular sport. Like boxing, it has strict rules

There was always great rivalry between the bands at Carnival, and most of them had fighters as well as musicians. When the stick fighters found the going tough, the bamboo players had the weapons to hand to pitch in. Many meetings between bands ended in street battles with bloody noses and sore heads.

Once again, therefore, the authorities clamped down. In the 1920s, they said that no more than ten men in a band could carry sticks on the street. This amounted to a ban on Tamboo, for the sound that could be made by ten men was nothing like enough to get the carnival dancing. The stick fighters went back to their yards, and the music makers looked around once more.

15

Alexander's Ragtime Band

At first, the new carnival sound depended on ordinary instruments. A generation grew up that knew nothing of Tamboo Bamboo. They made their music with trumpets and banjos. Then, one Carnival Monday, an altogether new sound poured joyfully into Port of Spain. A mass of people, calling themselves *Alexander's Ragtime Band* after a popular song of the time, made a tremendous sound and a compulsive rhythm. Pete Simon describes the scene:

> All sorts of improvised percussion instruments had been co-opted by this band—buckets, dustbins, pitch-oil pans, soapboxes, motor-car hubs, and other noise making devices. Led by Lord Humbugger, a tall commanding figure dressed only in a top-hat and long black overcoat, it made a startling impact. Everybody jumped in, passing bands quickly swelled the numbers, and when it reached Frederick Street, the main thoroughfare, nearly the whole city was jumping up to Alexander's Ragtime Band. The clear metallic ring of steel was the dominant feature of its sound.
> *Trinidad and Tobago* (ANDRE DEUTSCH 1975)

Alexander's Ragtime Band appeared in 1937, and at once the rhythmic instincts of the Trinidadians were awoken. The harsh sound of clashing steel seemed to be what they were searching for. At once, the inventive spirits got to work. At carnival the following year, every band was a **steel band**, and rhythm and noise was back.

Rhythm and noise, but no melody. At first, the exciting sounds and rhythms of the new steel bands were enough. But soon, the first **pan men**, as they called themselves, wanted to play tunes. But how do you get tunes out of dustbin or a hubcap?

There are many claims as to who was the first man to achieve this. Neville Jules, Winston Simon and Ellie Manette are amongst those most often mentioned. Probably, the truth is that they all contributed to the great discovery. There are enough stories about it, and one that is still heard concerns Winston ('Spree') Simon.

▼ A collection of items that early ragtime bands might have used as percussion instruments

Spree Simon

The legend is that one day Spree Simon found a dent in his dustbin. The dent was no doubt the result of a game of cricket in the road, or a stone thrown by some boy who never knew the contribution he was making to musical history. Spree Simon took a hammer to knock out the dent, and suddenly became aware that each blow with the hammer produced a different sound as the shape of the dent changed. Spree now used his hammer to add dents of various shapes and sizes, and, after a lot of experiment, he made a scale of notes.

No one is sure what was the first tune Spree played. Some say it was 'I am a warrior'. There is even a story that it was 'Mary had a little lamb'. It is enough that there was a tune.

For one reason or another, dustbins were not found to be ideal. They became hard to find. The sound was not satisfying to the perfectionists. the metal was thin, and the dents did not keep their shape for long; they had to be frequently retuned.

The solution was at hand. The war came and went, and to Trinidad it brought the American forces. The **oil drums** left abandoned by them were perfect materials for making pans. There were plenty of them, and a continuous supply, as Trinidad is an oil-producing country. They were all the same size and shape, and produced tones that did not clash with each other. They were made of higher quality metal than dustbins.

The story of the steel band is in many ways a story of struggle. It grew up as the music of the poor, rather as jazz did in the USA. It was not in any way respectable. From the drum bands of a hundred years previously, through the bamboo players, and down to the first steel bands, there was a history of violent behaviour side by side with the music. There is no doubt that the steel band yards were the meeting places for the hooligans and villains of the time. As with the Tamboo Bamboo bands,

▲ Steel band music started out as the music of the poor; there was often a lot of fighting as people crowded in the streets at carnival time

rivalry between steel bands ran deep—far beyond their music. Massive battles were not unusual. None of the players were considered musicians. They could not read music, and learned everything by ear. (This is still true, though nobody doubts their musicianship now.) The authorities still had a fear of primitive rhythms, and all that went with them. It was hardly surprising that just to belong to the steel band was enough to brand a youngster as a potential criminal.

In time, however, the steel band came to be recognized for its worth. The Trinidad government, after independence, encouraged major companies to sponsor bands. Even the Trinidad police started a band of their own. Pan began to be socially acceptable.

Carnival

Nowadays, steel bands can be heard all the year round and in many parts of the world. But the carnival in Port of Spain, Trinidad, remains the highlight of the steel band year. The carnival proper lasts for two days; preliminaries take place in the two weeks before. For some band leaders, the preparations for the carnival begin a year in advance.

In the weeks before Carnival starts, the mood is prepared by such events as the calypso fiesta, the choosing of the carnival King and Queen for the year, and, of course, the eliminating competition for the steel band of the year. About fifty steel bands enter, and each one has to play a calypso tune on the move. The best fourteen are chosen for the semi-finals, called **Panorama**, where they share the occasion with the hopeful Queens of the Carnival. Seven steel bands are selected for the finals on 'Dimanche Gras'—the Sunday before carnival. All the competition finals take place on this night.

In Carnival, the word 'band' means much more than just steel band. Each band (which may consist of a thousand or more people) works to a theme, and parades a gigantic masquerade. The band leader will decide that his band will represent, say, clowns, or devils and demons (these are two popular themes). A vast amount of energy is then poured into making the most fantastic costumes. These costumes, which are built on to wire frames for support, are sometimes four metres high or more. They have the most detailed stitchwork imaginable, and take weeks to make.

On the opening Monday of Carnival—called 'Jour Ouvert' or open day—everything starts at 4 a.m. In Independence Square at Port of Spain, the mayor makes his opening speech, a rocket is fired, and the long day begins. All the steel bands move in, and although there are stands for those who just want to watch

▼ This magnificent bird costume is more than twice the height of its wearer

18

it seems that most people join in, just because they can't resist it.

On this day, the main costumes do not appear. Even so, there will be plenty of fancy dress about, mostly humorous, often imitating and teasing well known personalities. Mostly it is a day for joining in. Each steel band gets a crowd of followers. They dance if they can, but the crowd is packed so tight that mostly they just get carried along.

The great day is Tuesday. On **Mardi Gras**, the great bands parade the streets. The colourful masquerade continues for hour after hour. Somehow, even in the weight of their elaborate costumes, the people dance. When their energy should be fading, a little more rum keeps them going. This almost endless 'jump-up' does of course end; on the stroke of midnight, the carnival is suddenly over, and Port of Spain returns to normal for another year.

Questions

1 Why did the character of the Trinidad Carnival change after the year 1833?
2 Why did the authorities fear drumming and tamboo bamboo?
3 Why was *Alexander's Ragtime Band* important in the development of the steel band sound?
4 Why did the players become dissatisfied with the first steel band sounds?
5 How did Spree Simon make a scale of notes on his dustbin?
6 What problems were suffered by the early steel band players?
7 Why do you think the 'Panorama' competition has this name?
8 What is a Carnival Band?
9 What are the French names for the Sunday, Monday and Tuesday of Carnival? Why do you think the names are in French?

Projects

1 Take a theme for a carnival band, and list all the characters and creatures that could be included in the band under this theme.
   Suggested themes: Clowns; Devils and demons; Pirates. Better still, think of one yourself.
2 On a large sheet of paper, design a costume or mask for one of your characters. This should be in two parts:
   a) The structure; how it is kept rigid; how it is attached to the wearer.
   b) The artistic design; how it is going to look.
   If your school or neighbourhood is having a carnival, you could make up the costume and wear it.
3 Make a collection of pieces of metal which make distinctive sounds. Work out how they can be held without deadening the sound. You now have the material for a very primitive steel band.
4 The struggle experienced by the early steel band players was similar to that of the New Orleans Jazzmen. You may be interested in finding out more about them, and making comparisons. The story of trumpeter Louis Armstrong is typical, and interesting.

Bass pan player (see page 23) ▶

3 The pans, and how they are made

Ordinary oil drums like the one shown in this picture are the basic material of the steel band. It seems an odd source of sound, and this may be the reason that there are still many musicians who do not take pan music seriously.

Nevertheless, the science of sound applies as much with a steel drum as it does with a gleaming and expensive kettle-drum. In brief, the vibrations made by striking a percussion instrument give off a wave motion detectable to the ear. These waves are known as 'frequencies'. If you hit a rough piece of metal with a stone, the frequencies will be confused, and the sound ugly. On the other hand, a smoothed and tempered piece of metal can be made to give off a sweet and resonant sound. If its shape is altered, the frequencies will change, and the note will be different. A steel pan has, in effect, a whole series of smoothed and shaped surfaces, each of which gives a different note.

At first, pan makers were pleased to be able to get a few different notes on to the top of an oil drum, and they did not worry very much how the notes were arranged. As pans became more sophisticated a standard arrangement had to be determined, otherwise players would have to learn each piece anew if given a different pan. There is still no international standard layout, but one has now been adopted in Britain, and education authorities and pan makers are coming into line with this.

The pan diagrams in this chapter conform to these standard layouts.

Oil drums like this one are the basic material of the steel band ▶

Bass pans

The biggest drums in the steel band are naturally the **bass**. This consists of five full-sized oil drums arranged in a horseshoe shape. They have a range of $1\frac{1}{2}$ octaves, from D to G. The lower end of the oil drum is cut away, like an opened tin, and the rim rests on a stand which holds it just off the ground. The part of the stand which touches the drum is padded or damped with leather or rubber, to prevent unwanted vibrations from spoiling the sound. For the same reason, the drums must not be allowed to touch each other.

Bass Range

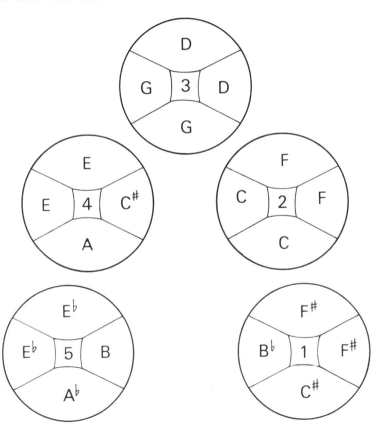

Cello pans

Next in size to the bass is a set of three pans called the **Cello**. These vary in size, but should ideally be half to threequarters the size of the complete oil drum. The cello and all the smaller pans are suspended on stands. As they are separate, the player can arrange them to suit his own convenience. There are seven notes on each cello pan, giving a range of nearly two octaves, from B to G. These pans are used to add notes to chord accompaniments, and sometimes duplicate the bass, one octave higher.

Cello Range

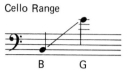

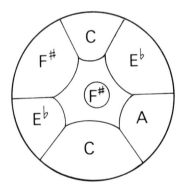

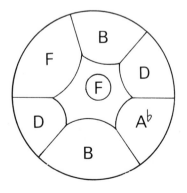

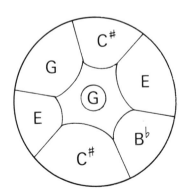

Guitar pans

The **Guitar** pans are next. They look similar to the cellos, though they are generally less deep. There are two pans in the set. Like a guitar in a dance band, these pans have the job of playing chords. They are sometimes known as strummers. Of course the player can only play one note with each hand, and the chords are normally filled out on the cello or double second pans. The guitars are fairly deep and mellow in tone. There are sixteen notes—eight on each pan—and they range from D to F.

Guitar Range

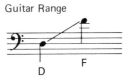

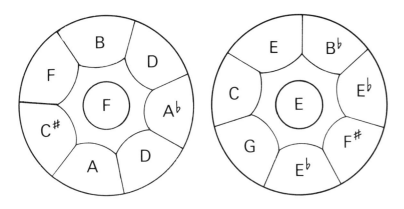

Double second pans

The **Double Second** pans are a pair similar to the guitars. They have more notes to each of the pair, a more ringing tone, and a two octave range. They reach a fairly high F♯. They can be used as strummers, playing chords like the guitar pans, and adding a further two notes to the chord in the whole accompaniment. However, their tone and range mean that they can be put to many more uses than this. Sometimes, they are given the melody line. More often, they play a harmony or counter melody, which they can sustain against two or three 'Ping Pongs' (see facing page). This greatly enriches the total sound of the band.

Double Seconds Range

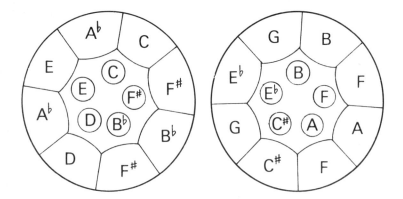

Tenor pan

The **Ping Pong** or **Tenor** pan is the highest in pitch. In fact, there are various kinds of ping pong, each with minor differences. The one illustrated on this page is known as a tenor, and is the kind most often found in schools in Britain. The name tenor is rather misleading, as it suggests a much lower range than this pan in fact has. It has thirty notes, giving an enormous two and a half octave range. More than any other pan, this remarkable instrument demonstrates the art of the pan maker as complex and delicate.

It is the lead instrument in a steel band, and even the smallest band will normally have two or three. They normally play the melody, working together. Sometimes the notes are unmarked, but experienced players glide over them effortlessly, adding improvisations and decorations to familiar tunes.

Tenor Range

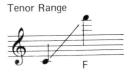

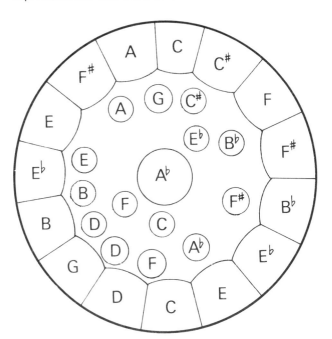

Making steel drums

The job of making a steel drum is a big one. It takes an expert a whole day; someone making their first one might spend a week and still not get it right.

A youngster interested in making one for his or herself *must* get the help of an adult. Certain parts of the work can be dangerous and should not be attempted without adult assistance.

It is heavy work. You would probably get blisters.

It is noisy work, with lots of hammering. You need to make it in a place where neighbours will not be disturbed.

If you are not put off by these difficulties you will find detailed instructions on the next few pages. If you are, then the account at the top of each page will be enough to tell you the main stages by which the job is done.

Sinking the pans

The first stage is **sinking**. This means hammering the end of the drum so that it becomes a smooth concave basin (the surface on to which the notes are put). The highest pans—the ping pongs—are sunk the deepest, and the depths become less until you come to the bass which is only sunk a small amount. This probably has to do with stretching the metal tighter to gain higher notes, just as you would with the strings on a guitar or violin.

◀ Guitar pans at carnival

Preparing for the job

YOU WILL NEED:

A 45-gallon oil drum (standard size). If you can't beg one try 'drums and kegs' in the yellow pages of the 'phone directory. The ends should be fairly free from dents and rust, and made of 18 gauge metal.

A sledge hammer or heavy club hammer

A smaller hammer

A punch

A cold chisel

Protractor, rule, chalk, compass, pencil

Fuel for a hot bonfire—traditionally a rubber tyre is used, but it gives off very smelly smoke; you will be better—and more popular—with wood.

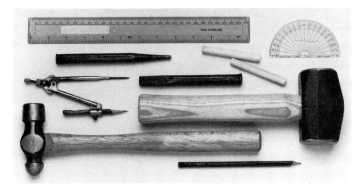

You may encounter difficulties not mentioned in this book. There is an instruction manual by Peter Seeger (*Steel Drums: How to play them and make them*, by Peter Seeger; Oak Publications, New York, 1964) which is detailed and very helpful, giving some alternative methods. I would certainly recommend this book—but be warned: he does not use the standard layout which was adopted after his book was published.

How to sink an oil drum

Start work with a sledge hammer, and beat very evenly from the centre towards the rims. You must take the greatest care to avoid making sharp dents. Use the full face of the sledge hammer—never the edge. If you do get a dent or crinkle, smooth it out at once with the smaller hammer. Hold your hands well apart on the sledge hammer, and mind you do not trap a hand against the rim of the drum. You will find the job easier if your shorten the handle of the sledge, or even use a heavy club hammer.

Work towards this shape:

Not this ⊢■⊣ or this.

Different pan-makers work to different depths. As a guide, I suggest the following measurements, from the level of the rim to the centre of the pan:

Ping Pong 170 mm

Seconds, guitars, etc. 100–120 mm

Bass 70 mm

To get it right, hold a stick marked with the correct measurement against the centre of the pan, pointing straight up. With your eye against the rim, check whether the opposite rim comes into line with your mark, as shown.

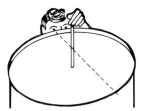

Now take your small hammer, and, using your other hand to feel out irregularities, gently tap them all out until you have a perfectly smooth dome. Your drum is then ready for the next operation.

Marking and grooving

The next stage is one where exact care is needed. This is **marking** the positions where the notes will be put, then separating them out by making grooves. Working from the centre, the maker draws 'spokes' to exact angles to the rim, then measures in the notes in their right positions. Then the pan has to be **grooved**. The job is done with a hammer and punch. The reason for it is to isolate the notes from each other.

Grooving – that is, marking the identations which separate the notes of a steel drum ▶

▼ Marking the position of the notes on a steel drum

How to mark and groove a Ping Pong

The ping pong is the pan with the most notes, and is thus the hardest to make. It is also the one to start with, being the pan the band most needs.

The first step is to find the centre. If the drum is 57 cm in diameter, take a piece of string just over half that length (to allow for sinkage) and hold a chalk at one end. Press the other end on the rim and draw an arc near the presumed centre.

Repeat this from two different parts of the rim. The spot where the three arcs meet will be the centre. When you do it you will be lucky if the three lines meet exactly. You will be able to judge the centre accurately enough for the small triangle or 'cocked hat' they make.

Protractor

From the centre you must now draw the radial lines to the rim, in chalk.

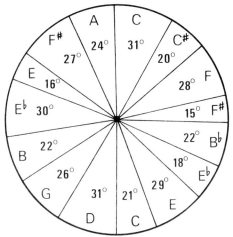

Copy this diagram on to the surface of the drum. You must use a protractor to be sure that the fifteen angles are accurately measured.

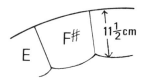

From the rim you now measure back $11\frac{1}{2}$ cm along each line and mark these points. Curved lines linking these points now complete the patterns for the notes round the edge, which may be pencilled in ready for grooving. The note names given in diagram 4 show which note comes in which angle.

The next stage is to mark in the round notes in the centre of the pan. With these, the most important thing is to get the circles the right size, as shown in the table below. They can be marked in by eye, with reference to the diagram on page 27. Do this in chalk—you can pencil them in when you have them evenly spaced.

Use a compass set to the following radii for each note.

A 110 mm	Ab′ 70 mm	C″ 55 mm	E″ 45 mm
D′ 97 mm	A′ 70 mm	C#″ 55 mm	F″ 40 mm
F′ 80 mm	Bb′ 65 mm	D″ 50 mm	F#″ 40 mm
G′ 75 mm	B′ 65 mm	Eb″ 50 mm	

When the pan has been completely marked out in pencil it is time to start the grooving. It is a good idea to turn the drum over and practise on the end you are not using. Only when you are sure you can make a really straight groove, and that you have the weight of the hammer blows judged to perfection, should you start the real job. It is not difficult to punch a hole through the already stretched metal, and you could spoil the hard work already done (although the odd small hole is not too great a disaster). On the other hand, if the groove is not deep enough, it will not do the job of isolating the notes from each other. If you are using a new punch, file it down a bit to round off the corners—sharp edges tear the metal.

The punch holes must be in line with each other. If the groove is not straight, it will be impossible to tune the note. Using your pencil mark, place the centre of the punch on the line, and punch a row of hollows about 1 cm apart. Punch the next set halfway between these hollows, and continued closing the gap until you have a continuous line.

Cutting, ponging up, and burning

When the pan is fully marked and grooved, it has to be cut down to the required size; only the bass pans use full-sized drums. Holes are bored in the rim for the carrying straps, and the pan is **ponged up**—that is, each note is hammered up from below into a shallow dome. This has to be done carefully; the surfaces, although domed, must remain smooth. Next, the pan is placed face down on a hot fire for a while; when it is really hot the steel is tempered by cold water. It is then ready for the most delicate and difficult task facing the pan maker—the **tuning**.

▼ Cutting the drum to size

▲ Ponging up—that is, hammering the notes from below into a shallow dome

▼ Burning the pan to help it to achieve a better tone when tuned

Cutting the drum to size

There are three ways of doing this job. It can be done with a hacksaw, but this takes a very long time. You can use a sharp cold chisel with a hammer. Perhaps the best way is to take your drum to a car repairer with an oxy-acetylene torch, and ask him to do it for you.

Whichever way it is done, the first stage is to mark in a line, the right distance from the rim of your pan, to ensure a straight cut. The ping pong should be cut to about 200–250 mm (enough to ensure that the belly, which is very liable to damage, does not stick out). The double second will be about 300 mm deep, the guitars about 400 mm. A nice size for cellos is about 700 mm, but they are made smaller for easy carriage, and so that they fit onto the regular stands.

If you are using a chisel to do the cutting, sit astride the drum and hold it an an angle from the metal, pointing away from you, of course. This method will take you about half an hour.

The strap holes should be placed where the rim meets the surface of the drum. Imagine nine and three on a clock face, and place the holes slightly up towards ten and two, so that the pan will hang at an angle towards the player.

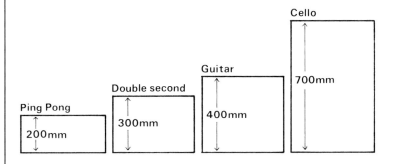

Ponging up

Turn the pan upside down, and work on each note with the small hammer. By tapping at the centre, not the edges of the note, you will raise it by about a centimetre. Then, when the time comes to tune it, you can do so by tapping down. The drawing across a section of the pan's surface shows the shape you are aiming for. It will be smoothly curved; free from dents. Your oil drum will at last begin to sound like a pan as you do this job—but it is not yet time to start tuning.

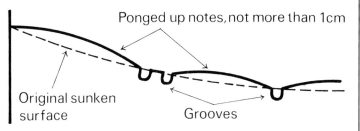

Tempering the pan

You now make a hot fire. When it is going well, put the pan face down on to it, and leave it for a while. Every pan maker will tell you a different length of time to burn it, and they all seem to work all right! As a guide, allow the paint to be burned off, and the metal to take on a greyish colour.

The hot pan must now be lifted away from the fire—use tongs, or sticks—and cooled rapidly. Have a bucket of water ready, and pour it in immediately. Watch out for hot splashes.

When it is cool, empty out the water. It is time to start tuning.

Stands

To tune the pan conveniently, it will need a stand. This must have a firm base, and jaws which are wide and deep enough to allow the pan to swing freely. It can be made from wood or metal pipe. Two suggested designs are shown.

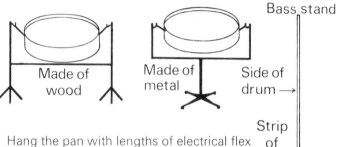

Made of wood

Made of metal

Bass stand

Side of drum →

Strip of tyre

Hang the pan with lengths of electrical flex through the strap holes. Strips of an old rubber tyre bolted to the bottom of a bass in three places will suffice as a stand to hold it off the floor.

▲ Tuning a pan is a very skilled process; electronic devices are often used nowadays

Tuning

The last job in making a pan is one that requires great patience. The tuner taps each note with a hammer until it is in tune. When he has done this for every note, he will have to start again, as most of the notes will have gone out again. Tapping down will usually bring the note down in pitch—but it does not always work out that way. When the pan has been tuned, it has to be handled gently. Quite small knocks can put it out again—and there are not many tuners around.

How to tune a pan

The mysteries of tuning will not be revealed to you in one page of a book. These instructions will tell you what to do, but the job seldom goes according to plan, and you must experiment your way past each difficulty. Do not be too disappointed if your early efforts fail. The failure can teach your more than any book.

You need your small hammer, a stick for playing (see page 46) and some pitch pipes. A melodeon, which gives you every note, is ideal.

When you ponged up before burning, you should have taken the notes to a pitch higher than they will finally be. To tune them down, you go the opposite way. Using the hammer and stick alternately, you will hear the note come down and down in pitch. Start with the lower notes on the ping pong.

Quite often, before it gets low enough, the pitch starts rising again. When this happens, turn the pan over, pong up again, and have another go. If you still can't get the note you want, move on to another note, and try the first one again later.

Most of your hammer taps will be at or towards the centre of the note. As before, you must keep the surface even and free from dents. If the sound of the note is imperfect, it may be because ridges and dents in the surface are causing dissonant overtones. Carefully pong up again, working out the irregularities as you do so.

Some of the smaller notes may be lower in pitch than you need after ponging up. These will be tuned by ponging up further, or by tapping *down* round the edges of the note.

When you have done every note, you will have to start again, because those you tuned first will have gone up in pitch as you worked on the others. You go round over and over again. Each time the differences get smaller—but you have to be patient.

A word of encouragement if you are looking over your finished pan. They generally get better with playing!

The Rhythm section

Some steel bands rely entirely on the pans, and add no other instruments. It is quite usual, however, to add a rhythm section to your band. If you want an authentic steel band sound, the instruments for this must be carefully chosen.

The ideal rhythm instrument is a pair of **conga drums**. Good congas are very expensive, and you may have to content yourself with more modest tom toms at first. Peter Seeger's book gives instructions for making a steel conga, damped underneath with chewing gum so that it gives a dull two-tone thud.

Maracas, **claves**, a **guiro** and a **cowbell** are very suitable. It is a temptation to put the band's less good musicians on to these intstruments, but beware; they make a lot of sound, and a rhythm instrument in an unsure hand can ruin the band's performance. All these instruments can be bought, but they can also be made very easily. Dried peas in two squeezy bottles

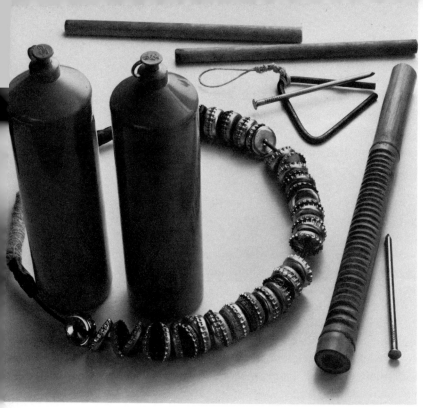

▲ Some home-made percussion instruments

make a good pair of maracas. Two sawn-off lengths of broomstick about 20 cm long, lightly held, and banged together are excellent claves. A piece of thick bamboo, notched and scraped with a nail is a guiro, and a car hubcap banged with a piece of wood is just as good as a cowbell.

Some quieter numbers sound good with **Indian bells**. A triangle gives a similar sound. These instruments again have a piercing tone, and should be used sparingly; to highlight one beat in each bar, for example.

With this array of instruments, it is doubtful whether you need a **drum kit**. Most modern steel bands do have one, however, and a good drummer certainly drives the band. If you don't have a good drummer, you are better off without the kit.

Questions

1 How is a steel drum tempered?
2 What is the lead instrument in a steel band?
3 Which pans are used to play chords?
4 Why are ping pongs sunk deeper than other pans?
5 Note positions are determined by lines radiating from the centre of the pan. Which have the wider angles, low notes or high notes?
6 How can a pan maker practise the skills of hammering and grooving?
7 In general, does tapping a note down make it sharper or flatter? (N.B. It doesn't always work this way!)

Projects

1 The sound made by every instrument comes from some form of vibration. Find out how the vibration is created on each of the following instruments:
   a) Triangle   d) Flute
   b) Piano   e) Trumpet
   c) Violin
2 Draw a diagram of a piano keyboard, putting an arrow against middle C.
   Using brackets, mark in the range of each set of pans in a steel band. Find out which pans overlap in range.
3 On a large sheet of paper draw a circle 57 cm in diameter, to represent the surface of a steel drum.
   Using the measurements and angles given in this chapter, and referring to the diagram on page 31, make an accurate full-sized drawing of a ping-pong.
4 Make a collection of as many as possible of the rhythm instruments listed on page 35. Make your own where necessary. You can use them to make interesting rhythm accompaniments for songs or other music.

4 The arranger at work

As few steel band musicians read music, they depend on an **arranger** to sort out what each pan will play, how the tune is **harmonized**, and what the final sound of the whole band will be. Every band has to have its music arranged.

The picture on this page, and the description of the pans in chapter three, show the arranger's materials. Every arranger has his own way of working, but it might be planned out like this:

1 Hear a piece which he thinks suitable for his band.
2 Listen to it until he has a clear idea of how he wants it to sound on the pans.
3 Get hold of **sheet music** for the piece. (Some arrangers scorn to use any music—they just listen and listen until they have it right.)
4 Note down what each player will be playing—the notes of the tune for the ping pong players, the notes of the bass, and the notes to be played on the inner drums to complete the harmony.
5 Work out a suitable accompaniment for the rhythm section.
6 Teach it bit by bit to the players in the band.

A steel band player soon develops a retentive memory. He has to learn and remember the notes the arranger teaches him, the length of each note, and the 'feeling' of how his part fits into the piece as a whole. If he goes wrong, with no music in front of him, it is only this feeling, which is developed by listening, that will put him right.

In this chapter, you will see how the arranger gets to work on several different tunes. The arrangements are simple—the sort of thing a band of beginners might play—and you would hardly need to read music at all to try some for yourself.

(title page) Steel band in a London school

Labelled diagram showing the position of the drums in the photo on the
▼ previous page

BASS

GUITARS AND DOUBLE SECONDS—HARMONY

CELLOS—HARMONY

PING PONGS—MELODY

RHYTHM SECTION

Michael row the boat ashore

This would be a good tune to teach a group of beginners. It is simple. It is well known. If you find it in a song book it may well be printed as below. The music is the **melody line**. This is given with the words underneath, and letter symbols over the top. The symbols are the **chords** which a guitarist would play to accompany the song. This example provides all the information an arranger would need.

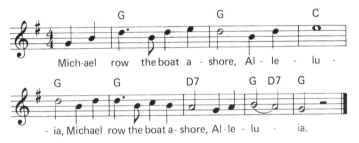

The ping pongs would play the **melody** as written:

count	1 2 3 4	1 2 3 4	1 2 3 4	1 2 3 4	1 2 3 4
play	G B	D ~BD E	D ~ B D	E ~~~~~	D ~~B D

count	1 2 3 4	1 2 3 4	1 2 3 4	1 2 3 4
play	D ~BC B	A ~ G A	B ~~ A ~	G ~

For the **bass** part, it will be adequate to play the **root note** of each chord (i.e. G G C G etc.), but both player and listener will soon find this boring, so it could be given some interest by playing the rhythm *taa-a-a-tate* in each bar:

count	1 2 3 4	1 2 3 4	1 2 3 4	1 2 3 4	
play	(rest)	G GG	G GG	C CC	etc

Some bridging notes could be added when the chord changes:

count	1 2 3 4	1 2 3 4	1 2 3 4	1 2 3 4	1 2 3 4
play	(rest)	G GG	G GG	C B A G	GG

count	1 2 3 4	1 2 3 4	1 2 3 4	1 2 3 4	1 2
play	G GG	d CC	d ~~~ D ~~~	G ~~	

Explanation of symbols used in diagrams
A ~~~~~~~ Rolled note   D Larger (lower) of the D notes on the pan
d Smaller (higher) of the Ds.

These simple devices give some idea of the ways in which an arranger can develop a piece in a personal way.

The guitar chords also give the arranger the basis for the **harmonies** played by the inner parts. A chord is a group of notes played together. While any such group of notes is a chord of sorts, there are certain notes which go together naturally to make the chords which are most commonly used. On the next page is a diagram giving almost all the chords that regularly appear in printed songs. They are arranged in groups according to the relationship between chords within the framework of a particular key. The notes which group together for each chord are given—it is a useful reference page for budding arrangers.

Michael row the boat ashore is written here in the key of G. The chords used most often in this key are G, D7 and C. From the diagram, you will see that the notes in the G chord are G, B, and E; the D7 chord is made up of D, F♯, A, and C; and the C chord consists of C, E, and G. These are the notes which the arranger will give to the double seconds, guitars, and cellos.

He decides to have the double seconds and the guitars working together on a **strum** to the rhythm taa ta-te taa taa. Each player will play one note of the chord on 'taa' followed by two together on 'ta-te' and 'taa'. This is a commonly used strum for these pans. The cello pans are to enrich the bass by playing **arpeggios** on the same chords. It works out like this:

chord	—	G	G	C	G
count	1 2 3 4	1 2 3 4	1 2 3 4	1 2 3 4	1 2 3 4
d/sec	rest	G gg G g / BB B	G gg G g / BB B	G EG G E / CC C	G gg G g / BB B
guitar	rest	D dd D d / BB B	D dd D d / BB B	E CC E C / GG G	D dd D d / BB B
cello	rest	G B D B	G B D B	C E G E	G B D B

Michael row the boat a-shore,   Al-le-lu   ia,   Michael

chord	G	D7	G D7	G
count	1 2 3 4	1 2 3 4	1 2 3 4	1 2 3 4
d/sec	G gg G g / BB B	D f#f#D f# / aa a	G gg D f# / BB a	g ~~~ / B ~~~
guitar	D dd D d / BB B	F# dd F# d / CC C	D dd F# d / BB C	d ~~~ / B ~~~
cello	G B D B	F# A D A	G B D A	B ~~~

row the boat a-shore,   Al-le-lu   ia.

key	key signature	tonic chord	dominant chord	dominant seventh	sub dominant	relative minor
C		C / C E G	G / G B D	G7 / G B D F	F / F A C	Am / A C E
D		D / D F# A	A / A D# E	A7 / A C# E G	G / G B D	Bm / B D F#
E		E / E G# B	B / B D# F#	B7 / B D# F# A	A / A C# E	C#m / C# E G#
F		F / F A C	C / C E G	C7 / C E G Bb	Bb / Bb D F	Dm / D F A
G		G / G B D	D / D F# A	D7 / D F# A C	C / C E G	Em / E G B
A		A / A C# E	E / E G# B	E7 / E G# B D	D / D F# A	F#m / F# A C#
Bb		Bb / Bb D F	F / F A C	F7 / F A C E	Eb / Eb G B	Gm / G Bb D

Come back Liza

We will look now at another very simple tune. This one has a real West Indian feel about it, and it is familiar through Harry Belafonte's gentle lilting recording. While it is easy to arrange and to learn, it has some interesting points of rhythm. The band will really need to listen to each other if they are to hold it together.

This tune has a lot of character, and the inner parts will point this up with a very light accompaniment with plenty of rests. It helps a lot to get a rhythm right if the player thinks of some appropriate words as he plays. In this case, the players on a double second, guitar and cello could mouth the words 'ev'ry time (rest) ev'ry time (rest)' over and over, for that is the rhythm they will be playing.

The bass player will anchor it down by the three notes of the chords once in each bar, to the rhythm 'Come . . . Liza, Come . . . Liza'.

Ever - y time I'm a-way from Li-za, Wat-er come to me eye.

Wat-er come to me eye.   Come back Li-za, come back girl,

Wipe the tear from me eye.   Wipe the tear from me eye.

The rhythm section can play a soft rumba rhythm, counting the eight half beats in each bar. There will be an accent on the first, fourth, and seventh of these half beats, pointed perhaps with a crisp penetrating sound such as a wood block.

The ping pongs will play the melody, and the whole band will keep it soft and smooth. The beauty of this tune, however, is that if the mood takes you, you can double the speed, and play it as a fast rhythm number just as effectively.

Count	1 2 3 4	1 2 3 4	1 2 3 4	1 2 3 4	
Chord	G	G	D7	G	
Ping Pong	D DD D D	D G B D	A BC DE D 〰		FIRST TIME
			A BC BA G 〰		SECOND TIME
Double sec.	$\frac{g}{B}$ $\frac{gg}{BB}$ rest	$\frac{g}{B}$ $\frac{gg}{BB}$ rest	$\frac{f\#}{a}$ $\frac{f\#f\#}{aa}$ rest	$\frac{g}{B}$ $\frac{gg}{BB}$ rest	
Guitar	$\frac{d}{B}$ $\frac{dd}{BB}$ rest	$\frac{d}{B}$ $\frac{dd}{BB}$ rest	$\frac{d}{C}$ $\frac{dd}{CC}$ rest	$\frac{d}{B}$ $\frac{dd}{BB}$ rest	
Cello	$\frac{G}{D}$ $\frac{GG}{DD}$ rest	$\frac{G}{D}$ $\frac{GG}{DD}$ rest	$\frac{A}{C}$ $\frac{AA}{CC}$ rest	$\frac{G}{D}$ $\frac{GG}{DD}$ rest	
Bass	G — B d	G — B d	A — d D	G — B d	

Ev'ry time I'm a-way from Liza, Water come to me eye ———

Count	1 2 3 4 5 6 7 8	1 2 3 4 5 6 7 8	1 2 3 4 5 6 7 8	1 2 3 4 5 6 7 8
Maraccas	♫ ♫ ♫ ♫ etc			
Congas	♫♩ ♩♫ ♩♫♩ etc			
Wood Block	♩ . . ♩ . . ♩ . etc			

Count	1 2 3 4	1 2 3 4	1 2 3 4	1 2 3 4	
Chord	G	G	D7	G	
Ping Pong	G G B D	GB B 〰	A BC DE D 〰		FIRST TIME
			A BC BA G 〰		SECOND TIME
Double sec.	$\frac{g}{B}$ $\frac{gg}{BB}$ rest	$\frac{g}{B}$ $\frac{gg}{BB}$ rest	$\frac{f\#}{a}$ $\frac{f\#f\#}{aa}$ rest	$\frac{g}{B}$ $\frac{gg}{BB}$ rest	
Guitar	$\frac{d}{B}$ $\frac{dd}{BB}$ rest	$\frac{d}{B}$ $\frac{dd}{BB}$ rest	$\frac{d}{C}$ $\frac{dd}{CC}$ rest	$\frac{d}{B}$ $\frac{dd}{BB}$ rest	
Cello	$\frac{G}{D}$ $\frac{GG}{DD}$ rest	$\frac{G}{D}$ $\frac{GG}{DD}$ rest	$\frac{A}{C}$ $\frac{AA}{CC}$ rest	$\frac{G}{D}$ $\frac{GG}{DD}$ rest	
Bass	G — B d	G — B d	A — d D	G — B d	

Come back Liza, Come back girl, wipe the tear from my eye ———

Greensleeves

The greatest charm of the steel band is the way it can be adapted to almost every kind of music. The great steel orchestras think nothing these days of tackling the *William Tell* overture, or Sibelius's *Finlandia*. Here, instead, is *Greensleeves*.

There is a no key signature to this music, because it has a **modal** flavour. Take care to play B♯ where you might expect a B♭, and notice that C♯ is sometimes varied with C♮.

The accompaniment is with rolled chords. If the band plays it nicely, it will be quite possible to close your eyes and imagine you are listening to an organ.

Count	1 2 3	1 2 3	1 2 3	1 2 3	1 2 3	1 2 3	1 2 3
Chord		Dm	Dm	C	C	Dm	Dm
Ping Pong	d	f⌇g	a⌇b a	g⌇e	c⌇d e	f⌇d	d⌇cd
Double sec.		f/A ⌇		C/E ⌇		f/A ⌇	
Guitar		F/D ⌇		C/G ⌇		F/D ⌇	
Cello		D/A ⌇		E/G ⌇		D/A ⌇	
Bass		d A	d	c G C		d A	d

CHORD GUIDE	etc. ⌇	A	Am	F
Double sec.		C♯/E ⌇	C/E ⌇	f/c
Guitars		A/C♯ ⌇	A/C ⌇	F/A ⌇
Cellos		A/E ⌇	A/E ⌇	F/C ⌇
Bass		A E A	A E A	f c f

Hey Jude

The final arrangement in this chapter, while still moderately simple, shows how the arranger can put his own ideas of how it should sound into practice.

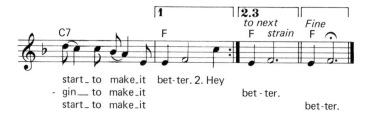

Ping Pong	C :A⌒⌒A c d	G⌒⌒GA	Bb f⌒⌒f e c	d c A⌒⌒c
Double sec.	C :F⌒⌒F	C⌒⌒EF	G⌒⌒Ab⌒⌒	A⌒⌒⌒F Eb
Guitars	:A/C ⌒⌒	E/G ⌒⌒	Bb/G ⌒⌒	A/C ⌒⌒
Cellos	:C/F ⌒⌒	E/C ⌒⌒	E/C ⌒⌒	C/F ⌒⌒ F C
Bass	:F F F F	C C C C	C C C	C F F F

Ping Pong	d/Bb d/Bb d/Bb gf efd	c⌒⌒FG A dc	c B A E	EF⌒⌒	EF⌒⌒
Double sec.	D⌒⌒Bb A	A⌒⌒F⌒ Bb	G	E⌒⌒	C⌒⌒	C⌒⌒
Guitars	Bb/F D/F D/F	Bb/F D/F	F A/C A/C F A/C	E Bb/G Bb/G	E Bb/G	A/C ⌒⌒	A/C ⌒⌒
Cellos	F/D d d	F/D d	F/C c c F/C c	C Bb/C Bb/C	C Bb/C	C/F ⌒⌒	C/F ⌒⌒
Bass	Bb F F	Bb F	A F F A F	G C C	G C	F⌒⌒	F⌒⌒

(Pause) ⌒

Ping Pong	EF⌒⌒	—F f eb d c c Bb	d⌒⌒⌒fd⌒⌒⌒f	Bb⌒⌒f d	c Bb
Double sec.	C⌒⌒	—F A G F Eb Eb D	F⌒⌒⌒F#⌒⌒	G⌒⌒⌒Bb⌒⌒
Guitars	A/C⌒⌒	REST	Bb/F ⌒⌒⌒	Bb/G ⌒⌒⌒
Cellos	C/F ⌒⌒	REST	D/F ⌒⌒⌒	D/G ⌒⌒⌒
Bass	F⌒⌒	REST	Bb Bb A	A G G F F

Ping Pong	c⌒⌒d c⌒⌒Bb	AG F⌒⌒⌒	—F f d d c c Bb	d⌒⌒f d⌒⌒f
Double sec.	G⌒⌒E⌒⌒	F⌒⌒⌒	—F A G F Eb Eb D	F⌒⌒⌒F#⌒⌒
Guitars	Bb/G ⌒⌒⌒	A/C ⌒⌒⌒	REST	Bb/F ⌒⌒
Cellos	E/C ⌒⌒⌒	C/F ⌒⌒	REST	D/F ⌒⌒⌒
Bass	E E C C	F⌒⌒⌒⌒⌒	REST	Bb Bb A A

Ping Pong	Bb f d⌒⌒c Bb	c d c⌒⌒Bb	AG F f c d eb⌒d	eb e# fg⌒⌒⌒C
Double sec.	G⌒⌒Bb⌒⌒	G⌒⌒E⌒	F⌒⌒F A Bb	C⌒⌒⌒Bb⌒⌒⌒C
Guitars	Bb/G ⌒⌒⌒	Bb/G ⌒⌒	A/C ⌒⌒ REST	Eb/C Bb/G ⌒⌒⌒
Cellos	D/G ⌒⌒⌒	E/C ⌒⌒	C/F ⌒⌒ REST	C/F E/C ⌒⌒⌒
Bass	G G F F E E C C	F⌒⌒REST	F F G G C⌒⌒

You could notice two things about this arrangement. First, the counter melody on the double second, which strays outside the notes strictly belonging to the chords given. Second, the inner parts vary between rolls and strums according to the character of that part of the melody. A little tension is added by beginning the second strain with ping pong and double second only—the accompanying pans join in with their rolls on bar two.

Questions

1. What is the difference between an arranger and a composer?
2. In what ways does a musician in a band (of any sort) make sure he keeps in time with the others?
3. What word describes rhythmic playing on both a guitar and a guitar pan?
4. What keys are announced by the following key signatures:

a)   b)   c)

5. Which chords are made up the by following notes?
   a) G♯ B E   b) B♭ F D   c) G A C♯ E
6. Which arrangement has a counter melody on the double second?
7. Which is in triple time (Three beats to the bar)?

Projects

1. Work out the notes which make up this series of chords, most of which do not appear on p. 40.

Key	Key sig.	Tonic	Dominant	Dominant 7th
A♭		A♭	E♭	E♭7

Sub dom. D♭   *Relative minor* Fm

2. *Come Back Liza* was made popular by the singer, Harry Belafonte. Listen to some of his many records.
   A modern Trinidadian calypso singer is 'The Mighty Sparrow'. See if you can get hold of a record of his.

3. In full the rhythm names in this chapter go like this:

Taa-aa Taa-aa / Taa taa taa taa / Ta - te ta - te ta - te ta - te /

Taa ta - te ta - te taa etc.   Tafatefe

Here are some phrases set to these rhythms:

Steel band

Taa - aa   taa - aa

Spree Simon

Taa - aa taa taa

Ragtime band

Taa taa taa - aa

Port of Spain, Trinidad

Ta - te taa ta - te taa

Make a list of the names of several friends, football teams, etc. and express their rhythms like this.

4. Try an arrangement of your own. You might be interested in this spiritual. It is a quiet, intense piece of music; will you accompany it with rolled chords, or a slow strum?

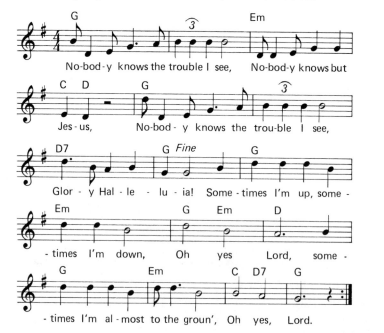

No-bod-y knows the trouble I see, No-bod-y knows but
Jes-us, No-bod-y knows the trou-ble I see,
Glor-y Hal-le-lu-ia! Some-times I'm up, some-
-times I'm down, Oh yes Lord, some-
-times I'm al-most to the groun', Oh yes, Lord.

5 Starting a steel band

Perhaps you are reading this book because you would like to start a steel band. The easiest part is to gather together an enthusiastic group of players, and it's a lot of fun learning together. However, you need time, money, and space, and you need an adult to get you organized. Perhaps your very first step would be to find yourself this sponsor.

Make a start by asking a teacher or youth club leader if they will help to get you started. He will be able to do a lot for you by organizing your accommodation and practice, and perhaps by getting the support of your education authority in finding and employing a tutor. The tutor must be a pan man—one who knows steel bands through and through, and can instinctively produce the right sound from your band. This is a very special quality; maybe part of it is born in him.

Assuming you can find a tutor to come to you regularly, you have another problem to solve before you get your pans. Where are you going to keep them? If possible, you need a special room for them, otherwise a lot of time will be spent in setting them up and putting them away when you could be practising. You must consider the other users of the building, too. A steel band is very noisy. The more isolated the room is the better. I know of a band which practises in the basement of a school, among the boilers.

The pans themselves are fairly readily available in this country, and some of the organizations listed on the inside back cover will advise you of a maker near to your area. A basic set to start off would consist of two ping pongs, a double second, a guitar, and a bass. This set will not be cheap—few musical instruments are—but all the same, I would advise you to start off with professionally made pans rather than make your own, unless you have an expert in your group. Specify the standard layout as shown in this book when ordering your pans.

(title page) A London school steel band playing Christmas carols in Trafalgar Square

Caring for your pans

You must look after your instruments. It is very easy to think of them as a load of old oil drums and treat them that way. If you take the pans out to play them somewhere, well-meaning people helping you to lift them need to be warned to handle them gently. Passers-by can rarely resist the temptation to 'have a go' using their fingers, or, worse, any convenient stick they can lay their hands on. This treatment can easily put your pans out of tune; retuning is difficult and costly.

You can put the drums out of tune yourself by banging them too hard, or by using unsuitable or worn-out sticks. Remember that the wooden part of the stick must never strike the pan, and that when the lining wears thin, leaving exposed wood, it must be renewed.

Making the sticks

Most pan men make their own sticks. The basic stick is a piece of dowelling about 15 cm long. Find a length which feels comfortable in your hand and suits you. The playing end is tightly wrapped round with a strip of rubber. This should be thick enough to have a little bounce when you hit the note, but not so thick that the sound is deadened. No precise amount can be given; it varies with different pans and players. Experiment until you find what is right for you. You can use a piece of innertube—for some reason red is better than black—or a piece of rubber that is used to wrap round cricket bat handles. Failing

▲ All the tools you need to make your own drum sticks

Playing your pans

When you start to play any instrument, a lot of basic skills have to be practised over and over, and mastered before you can make any real progress. The suggestions that follow will give some idea of these skills. Your tutor will demonstrate, and guide you.

The first thing you will have to learn is how to hold the sticks. As with all percussion instruments they need to be held rather loosely, so that they bounce off the notes and give an echoing tone.

One of the first things to practise is **rolling**, which is playing the same note over and over again in quick succession. This is the only way to play a long note on a steel drum, and you should try to play it so delicately that the repeated notes seem to merge into a single sound. Gradually get louder and softer. This is a very good exercise for stick control.

either of these buy some wide elastic bands, making sure that the rubber is supple, and not too thick. The proper way to finish off is by tucking the rubber into itself; however, if you find this difficult, don't be ashamed to use a little sticky tape.

The bigger the pan you are making the stick for, the more rubber needs to be wrapped round the stick. For double seconds, guitars and cellos in that order, you use more and more elastic. It is not uncommon to see the sticks for these pans given an outer layer made from a strip of foam stretched tightly round, and taped into place. Rather thicker dowelling can be used for these pans.

The bass sticks are usually made from a hard rubber ball with a hole to push the stick into. These can be quite heavy for a young bass player to control, and may be filed down like the one shown in the picture. Alternatively, a single ball can be sawn in half to make a pair of sticks.

Some **scales** must be learned, so that you can find your way around the notes on your pan. These have to be played time after and time, gradually increasing the speed as you gain confidence. It is very important indeed that each note is played with the correct hand. If you form incorrect habits early on, you will never be able to play fast runs or tunes. The notes for two useful scales are given, with left or right hand shown (for the tenor pan). Learn to play them forwards and backwards.

N.B. Small letters = higher notes

G major scale	G	A	B	c	d	e	f♯	g
	LH	RH	LH	RH	LH	LH	RH	LH
D major scale	D	E	F♯	G	A	B	C♯	d
	LH	RH	RH	LH	RH	LH	RH	LH

When you want a break from scales, try playing some **arpeggios**. Counting a steady 1 2 3 4, play this:
G B d B (four times) D F♯ A F♯ (four times) G B d B (four times) etc.

This is a particularly useful exercise on the double second and guitar pans. Learn to change from one arpeggio to the other without hesitating. When you can play it smoothly speed up a bit. After this you could work out another sequence of arpeggios to try, using the chord chart on p. 40.

All the above practice can be done individually, but of course a band has to work as a team, and you have to learn the skill of listening to the others and blending with them as you play. A good way to start this is with several members of the band practising scales in **unison**, with the aim of keeping together. You soon learn that the music won't wait while you correct a mistake.

You can never learn to play steel pans by reading a book. You'll just have to join — or start — a band. If that is what you are about to do, then good luck with it. You have a lifetime of thrilling music ahead.

Glossary

Alexander's Ragtime Band   Popular song in the 1930s. The name was taken by one of Trinidad's Carnival bands.

Arpeggio   The notes of a chord played one after the other.

Arranger   (Sometimes called the Captain) One who arranges how his steel band will play a piece of music.

Bass   Instrument or voice that plays the lowest notes. Tuba, bassoon and double bass are examples in the orchestra. The steel band bass is made from full sized oil drums.

Cannes brulées   (French) Cane burning. A plantation festival.

Carnival   There are carnivals everywhere — but the Mardi Gras Carnival in Port of Spain is exceptional, and people travel from all over the world to be there.

Cello   In the orchestra, the cello is a stringed instrument, just higher than the bass. Cello pans are the equivalent in a steel band.

Chord   A group of notes played simultaneously. On a piano, chords are played by pressing several notes together. On a guitar they are played by sweeping the hand across several strings.

Claves   Percussion instrument. Two pieces of hard wood are struck together.

Conga drums   A pair of drums giving two distinct notes. Played with the fingertips or palm.

Cowbell   Percussion instrument. Bell shaped piece of metal struck with a stick.

Double second   A pair of pans in a steel band which play just lower than the ping pongs.

Grooving   Making the grooves that separate the notes on a pan.

Guiro   Percussion instrument. Notched cane which is scraped with a nail or another stick.

Guitar   In a band, the guitar plays the chords that fit the harmony of the piece being played. In a steel band, the guitar pans do the same job.